NO NONSENSE VETERANS

OUT OF THE UNIFORM, BACK INTO CIVILIAN LIFE

An invaluable resource for veterans seeking clear, actionable guidance to navigate the often bureaucratic nightmare complex landscape of VA benefits and assistance programs.

LT. COL. JASON G. PIKE, USA, RETIRED

ISBN Paperback: 979-8-9889610-1-7
ISBN Electronic: 979-8-9889610-0-0

Library of Congress Control Number: 2023915777

Portions of this book are works of nonfiction. Certain names and identifying characteristics have been changed.

Printed in the United States of America.

Lt. Col. Jason G. Pike, USA, Retired
JasonPike.org

DISCLAIMER
Although this publication is designed to provide accurate information regarding the subject matter covered, the publisher and the author assume no responsibility for errors, inaccuracies, omissions, or any other inconsistencies herein. This publication is meant as a source of valuable information for the reader; however, it is not meant as a replacement for direct expert assistance. If such a level of assistance is required, the services of a competent professional should be sought.

TABLE OF CONTENTS

Introduction v

A Quick Overview of The Veterans'
Administration.. 1

First Things First: General Eligibility.....................................9

How This Book Is Organized..17

Applying..21

Healthcare...25

Va Education Benefits..37

Va Home Loan And Housing Benefits..................................45

Transitioning Into Civilian Life 49

Moving On: Va Burial & Memorials.......................................55

Mental Health Care 57

State Benefits...59

A Whole World of Opportunities...61

Introduction

"Jake is the only person I know who took a cut in salary when he graduated college..."

-Dennis Earl Pike

A S I WAS GIVING INITIAL THOUGHT TO THIS BOOK, I asked myself what qualifications I had to take on a project like this, to write a self-help manual on how to navigate the many benefits programs offered by the Veterans Association and many others, both private and state. The conclusion that I reached was that besides a particular attitude of mind, I had no qualifications at all. But an attitude of mind is often the difference between shaking the tree and settling for just the low-hanging fruit.

Where did that attitude come from? Well, growing up in the South, my Dad took a sink-or-swim attitude toward his kids. He always said that, at eighteen (nineteen, in my case, because I failed first grade), when you graduate high school, what happens is you're on your own, and he meant it. After that date, he would not even accept a collect call from any one of us. It was a harsh lesson, but it was a great and valuable one, too... and I took it literally because he talked about it all the time. As a kid, I cultivated and sold pot and traded beer and liquor in the schoolyard that I got older guys to buy for me and all kinds of other stuff. I knew the value of a dollar, and I was not afraid to stretch the rules to get one. My Dad looked at this and thought maybe I should be in business, something like a bait and tackle shop. "Son," he told me, "you have three choices: you can wear a set of coveralls, a suit and tie, or a uniform; it's up to you."

I had bigger ideas than a bait and tackle shop.

I guess if you are reading this book, then you have probably already read my memoir (if you have not, there is a link below), and you will likely know that, throughout my life, I have dealt with a pretty severe learning disability. Failing first grade seemed to set the tone for how the school and local education board pictured my future. I was encouraged *not* to contemplate college at every turn, which I guess was a pretty fair approach considering the evidence before them. I was not college-bound; that much was clear, and I had no business aspiring to anything more than a blue-collar career.

Then, one day, I went to the National Guard recruitment office and raised my right hand. At that moment, the clock started ticking on my military career, and against all expectations, I enrolled at Spartanburg Methodist College, where, by the way, I also signed up with the ROTC. With this came some financial support in the form of the Montgomery GI Bill, but there are limits to that and I needed more. There was a TV ad going around at the time featuring a goofy character called Matthew Lesko whose angle was to dress up in colorful suits decorated with question marks to sell a book called *Getting Yours: The Complete Guide to Government Money*. The book was just a list of telephone numbers to inquire about and apply for free stuff from the government. There was even a section on free stuff for felons. I called up a few of those numbers, and although nothing much came of it, it got me thinking.

Like most colleges, Spartanburg Methodist has a financial aid office with a full-time counselor, and looking at that, I figured there must be something in it. I went in and asked, and sure enough, the information I gathered told me that there was a whole lot of extra funding out there. It was not easy for a kid like me from an intact and reasonably affluent middle-class family, and I needed to show a pretty good GPA to qualify. I remember a kid on the campus saying, "Man, the only way you get a 3.5 is to take a low load." He was right, and I did, which put me in a good zone to start picking up scholarships. I applied machine-gun style for every damn thing I could think of, and although I screwed up a bunch of applications and was turned down by many others, a few did hit the target. It may have been just a couple of hundred bucks here and there, but those dollars mounted up pretty quickly.

A lesson I learned early on was that a thank-you letter goes a long way and that it is very important to attend every donor's banquet. I had one scholarship taken away from me because I was working and could not make it to one of those banquets. The college administration wants all their wealthy sponsors to see the students that they support, that are doing well, and that their money is being well spent. I ignored that invitation because I had a job to go to, and that was the last time I heard about that scholarship; that five-hundred dollars disappeared real fast. After that, I went to every banquet; if I could not, I would always write a letter of apology. If you are picking up free money, don't disrespect the hand that is giving it out.

The bottom line is that although you may have one scholarship and your college is paid for, one scholarship can always lead to another. If you need more money, go and apply for it. You can have double or triple scholarships; sometimes, they just hand out a bit of cash. A few bucks here and there make a lot of bucks at the end of the day. Sometimes, it is not about scholarships or entitlements but just a talent for figuring out how to get free stuff. I once got free rent for mowing the grass once a month. I also had a knack for finding free food. Most hotels offer breakfast, and I would walk in and eat, leave, and occasionally take some food out with me for later. Sometimes, I would order pizza over the phone with a false name, usually something no one likes, such as anchovies or onions and vegetables. Then, about two hours later, I would call in and ask for any "bad orders," with me usually just getting what I ordered.

Now I hear you ask: what has all of this to do with the VA and tapping into VA benefits? Well, the answer is everything. There is stuff out there in the VA universe that you can't even imagine and getting into it is more about a mindset than any particular acumen or qualification. My Dad had a standing joke that he would tell all the time. He would say, "Jake is the only person I know who took a cut in salary when he graduated college." And he was right. When I was cut loose from all the benefits and endowments, I surely was a whole lot poorer. What it did do, though, was to cause me to look more carefully at what the Army had to offer, and like the college system, there is a whole lot to be had, particularly after retirement.

This book deals with some basic practicalities of navigating the VA but also helps you understand what you are entitled to and how to get it. I have often heard vets say that they were never in combat, they have no serious injuries or disabilities, and that the money is for those guys, the frontline combat soldiers, but not for me! Nothing could be further from the truth. One guy picking up extra benefits does not take anything away from another.

In the VA, disability ratings are pretty important. I don't like the word "disability" because it implies "inability," and when dealing with the VA, that is not true. Disability can mean anything. Hell, even a toenail fungus is regarded as a disability. If you happened one day to be playing nighttime basketball on base and put your back out, you are considered government property, which is a disability under the code of the VA. Even just the normal wear and tear associated with your daily life as a serving member qualifies you to be considered for a disability rating, which in turn affects the level of and access to VA benefits.

I understand that the VA is a vast, complex, and intimidating entity. Even credentialled advisors don't know everything. However, if I learned one thing in the Army, it was looking at the picture from the ten thousand-foot level. I never did master the left-right-left school of military thought or the simple righty-tighty-lefty-loosey business of hands-on soldiering, but I was always able to get the big picture. This is a big-picture look at the VA, what it has to offer, how to get what is yours. Although I have no special qualifications, I have never been afraid of a long shot, and I have had enough good results from long shots to say that it is always worth trying. What is the worst that can happen?

A Quick Overview of The Veterans' Administration

*"There is only one place for the Veterans of America:
in the Cabinet Room,
at the table with the President of the
United States of America."*

-George H.W. Bush

ANYONE WHO HAS READ MY MEMOIR WILL KNOW that the guiding principle of my life has always been the Five Ps, known in military parlance as "Pre-Planning Prevents Piss Poor Performance." The Five Ps is a simple principle which basically means the more diligently you prepare the ground, the better the result will be. One of the key elements of the Five Ps is knowing that ground intimately, so before we plunge into what is under the hood, let us look at the background of the Veterans Administration, what it is, and how it came to be.

Many things set the US military establishment apart, and one of those is the enormous effort put into establishing possibly one of the best, and certainly one of largest, systems of veteran benefits, care, and support in the world. No other military establishment even comes close. This was something that followed the United States into the era of independence, and it is something that subsequent administrations have all taken very seriously.

The history and development of the US Department of Veterans Affairs has shadowed the history of war in this country. The first of these, of course, was the Revolutionary War, the founding struggle for our nation, which resulted in something around 8,000 casualties. Since then, there have been ninety-nine wars, most notably the Civil War, the World Wars, the Korean War, Vietnam, and the various Gulf Wars. These wars collectively claimed the most lives and resulted in the most injuries and disabilities.

The current Veterans Administration is the result of a long and ongoing evolutionary process dating back to the events of the Revolutionary War when the Continental Congress, in an effort to encourage recruitment, offered pensions for any soldiers who might be disabled in the service of the new nation. The first Pension Office and its original four employees managed the system. At the end of the Revolutionary War, the office was renamed the Pension Bureau, and it misaligned with the War Department. It struggled for decades to handle the claims of disabled soldiers of the nation's first war. After 1832, the Bureau was replaced by a new and much more streamlined Pensions Office, functioning briefly within the Department of War until it was moved in 1849 to the Department of the Interior, where it was renamed Bureau of Pensions.

The Civil War was the first mass casualty event in US military history, and in its wake, the Bureau was greatly expanded. The aftermath of the Civil War also saw the emergence of numerous state veterans' homes, which were, in essence, care facilities for indigent old soldiers compromised in one way or another or made homeless as a consequence of service during the Civil War. Because domiciliary care was offered to all residents of these homes, medical and hospital treatment was made available regardless of whether injuries or diseases occurred as a consequence of war. Destitute and disabled veterans of the Civil War, Indian Wars, Spanish–American War, and Mexican Border periods, as well as discharged regular members of the Armed Forces, were cared for at these homes. However, by the first decade of the new century, many veterans of those old wars were dying off, and the Pensions Bureau was scaled back until the modern Veterans Administration eventually absorbed it.

WWI marked a radical turning point in the scope and organization of the Pensions Bureau, which emerged in 1914 in revised form as the Bureau of War Risk Insurance. The original function of this Bureau was to insure US ships and their cargo from the hazards to be expected once war broke out in Europe. In October 1917, as the US was preparing to enter the war, this relatively small government agency exploded in size and scope to handle not only the insurance of naval assets but also deal with salaries, benefits, and insurance for all American service members involved in the US war effort, as well as facilitate the care and treatment of disabled veterans. Thirteen months later, with the end of WWI, the agency was tasked with mustering out returning service members, evaluating claims for disability compensation, and making arrangements for medical treatment, care, and hospitalization.

Although responsible for most veterans' programs, the Bureau of War Risk Insurance did not have under its remit the rehabilitation of disabled veterans, which fell under the more general purview of the Public Health Service. As a consequence, for over two years after the end of WWI, veterans were required to apply for eligibility with the Bureau of War Risk Insurance before they could then apply to the Federal Board of Vocational Education for approval for a specific training program, resulting in a critically inefficient and inept system that was a source of enormous frustration to veterans.

All of this resulted in a lot of hand-wringing and head-scratching in the various departments of government until, in 1921, under the presidency of Warren G. Harding, Congress created the Veterans Bureau. This was the first government agency explicitly dedicated to veterans affairs, but it was imperfect and subject to significant internal corruption. More soul-searching followed, resulting in the World War Veterans' Act of 1924, an article of legislation that established many of the essential elements of the Veterans Administration as we know it today.

Perhaps the most interesting and formative chapter of the evolution of the VA was the Depression era and the prelude to WWII. Besides the World War Veterans' Act of 1924, Congress passed legislation to provide a "bonus" to WWI veterans to recognize what was initially a volunteer army that served overseas during an era of full employment and enormous wealth accumulation at home. This bonus, however, was not to be paid immediately, and most WWI veterans were given a certificate that could not be redeemed until 1945.

When the Great Depression descended on the United States, a significant number of military veterans found themselves destitute. Tens of thousands traveled to Washington DC and occupied the National Mall in the summer of 1932, camping out for weeks, demanding immediate payment of the bonus. President Herbert Hoover responded by deploying the police backed up by regular units of the Army to clear them out. The newspaper-reading public was treated to images of ragged and hungry veterans confronting police and soldiers armed with tanks, machine guns, and rifles. They were cleared out of their shanties on the Mall by force and driven off by tear gas, clubs, and bayonets. The lingering images of this shameful affair led to Hoover's defeat and the debut of Franklin D. Roosevelt.

Roosevelt took a long look at a patchwork system he had inherited before he set about reforming and modernizing it with a series of executive actions. In his first one-hundred days, he passed the Economy Act of 1933, abolishing at the stroke of a pen all of the post-Civil War veterans' benefits laws that had been in place to date. In a series of subsequent executive orders, he created what is often regarded as the progenitor of the current veterans' benefits system.

However, Roosevelt's authority to essentially unify a sprawling and unwieldy system also gave him the power to slash veterans' benefits to help pay for other New Deal priorities. His subsequent relationship with various veterans' bodies and organizations could have been better. While he tried to mollify them by giving veteran leaders prestigious appointments and assignments, he needed help to overcome solid, grassroots resistance to much of what he proposed. Ultimately, he was defeated by the Independent Offices Appropriations Act of 1935, which stripped him of much of the executive authority the famous Economy Act granted to him.

When the American Legion proposed establishing a package of benefits going far beyond the mainly educational application it is known for today, Roosevelt took a pragmatic position. He put his weight behind the development of the G.I. Bill. The journey of the Bill through Congress was not without disruption and acrimony—there were, for example, some congressional members who opposed such benefits being made available to non-white servicemen—but ultimately it passed, creating an entirely new paradigm of thinking across the spectrum of veterans benefits and care.

Harry Truman inherited Roosevelt's skepticism, and like his predecessor, he tried hard to devolve veterans affairs to various general bureaus as an adaptation, or so he said, to the Cold War. He got no further than FDR, checked by the American Legion and various other veteran organizations. It was under Eisenhower, a soldier and a man with field experience in WWII, that the Veterans Association consolidated and began to organize itself into what we know today as the VA. Eisenhower was probably the man most determined to break up the VA and return its functions to the general administration, which was part of his deep distrust of the industrial-military complex. He was an ideological figure and had seen the power of armies. Still, he, too, was unable to dislodge the monolith, and it was left to grow through several more wars, acquiring cabinet-level executive department status only under the administration of President Ronald Regan.

From an organization founded to provide pensions and a handful of other benefits to veterans of the nation's early wars, the Veterans Administration has grown into a massive organization aiding veterans and their families in a multitude of different ways. Among many other benefits, the VA provides healthcare (in fact, the most extensive healthcare system in the world), disability compensation and rehabilitation, education assistance, home loans, and national cemetery services. Among these and many other benefits is the key element of this book, so let's get into what the VA has to offer and how to get it.

PRO TIP

Think eBenefits! eBenefits is a joint VA and Department of Defense web facility that provides resources and self-service to veterans and their families to apply, research, access, and manage their VA and military benefits and personal information through a secure Internet connection. Check it out at www.ebenefits.va.gov.

First Things First: General Eligibility

"I calculate that I will receive over one million dollars in tax-free VA money if I live to be seventy-six and at my retirement at the age of forty-eight. My last year in the Army was my 'million dollar year' as it set me up with the VA, not to mention all the second and third-order benefits."

-Lt. Col. Jason G. Pike, USA, Retired

BEFORE WE GET INTO HIS CHAPTER, I would like to tag a story in my memoir with particular relevance. I have emphasized the point in a couple of different places that a dishonorable discharge is not necessarily the end of the road for a successful VA benefits application. Below this section, I dwell a little bit on the technicalities, but for the moment, I want to share how I had a very close brush with that reality.

One night, I was out in San Antonio, drinking heavily at the Midnight Rodeo, which is a country and western bar, and with what was at that time pretty typical recklessness and irresponsibility, I got behind the wheel of my truck and headed through town to the Taco Cabana to pick up some food. I was flagged by a cop under pretty strange circumstances. I was in the queue at the Taco Cabana when I saw the cop through the plate glass, motioning me to come outside. I figured he needed help or something; maybe he knew somehow that I was also a uniformed member, so I went outside, and the bastard sobriety-tested me, cuffed and stuffed me, and booked me for DUI.

That was the beginning of a major shit show in my life. I was in my mid-thirties, a captain, undergoing advanced training, and just beginning to date the woman who would become my wife. Things were so much on the up at that time that suddenly getting kneecapped by a DUI in the middle of it all just seemed so damned unfair. But I brought it on myself; there was no denying that, and I had no one to blame but myself. At the time, I was not thinking so much about how it would affect my discharge status as much as I was concerned about the impact it would have on my career. However, with a less-than-honorable discharge, I would certainly have struggled to get into many of the programs that were, in the end, available to me when the date of my separation came.

It was my Dad who observed that so much money and resources had by then been channeled into my training that it would make no sense for the Army to show me the door if I could prove that I was an indispensable asset. "Not so many bug men out there," he said, "that they can afford to throw one away."

I also got to talking one day to an NCO who told me that an obscure regulation allowed for a soldier's record to be restricted if it could be proven that he was thoroughly redeemed and too valuable to throw away. It was a long shot, but I got to work, and eventually, I was successful. The story is in my memoir, but the long and the short of it was that I went hog-wild with training and simultaneously gathered as much official and unofficial support as possible. Even Major General James A. Peake, the senior officer who issued my original GOMOR, put his signature to an appeal on my behalf, along with many other officers and even a state senator. Interestingly, General James Peake was confirmed as Secretary for Veterans Affairs on October 30, 2007.

Ultimately, I succeeded, and my record was moved to the restricted file. My career continued, and as the record will show, I retired after thirty-one years at the rank of Lieutenant Colonel, with an honorable discharge and access to the full range of benefits offered by the VA.

The moral of the story is *to work the Five Ps*, and where there is a will, there's a way! I have often found in the military that if you ask a different person on the same day or the same person on a different day, you will get a different result each time. If you keep trying, you will eventually get what you are looking for. The military, and by extension, the VA, is a vast, complex bureaucracy. Because of that, it has an impersonal complexion at an administrative level that can work both for and against you.

I once knew a soldier who beat up his commander and was thrown out of the Army after a court martial with a dishonorable discharge, only to later be rehabilitated by a discharge status upgrade when it was eventually understood that his combat experience had screwed with his head in a classic case of PTSD. It was an uphill struggle and it took time, but he was successful in the end. That is not always the case, particularly if you are guilty of something really egregious. Still, the bottom line is don't walk away from your benefits just because you have a less-than-honorable discharge. These things can be changed.

Now, the basic question is: who is eligible for VA benefits? According to the **Department of Veterans Affairs,** *"You may be eligible for VA health care benefits if you served in the active military, naval, or air service and didn't receive a dishonorable discharge."*

In other words, so long as you have in the past served in any branch of the United States uniformed services, the door is open to you. Some benefits require war-time service, and as a rule, dishonorable and bad con-duct discharges, issued as a consequence of general courts-martial, are a disqualification for VA benefits. Incarcerated veterans are also generally disqualified, and veterans subject to outstanding felony warrants and their dependents are disqualified regardless.

However, if you were subject to a dishonorable discharge for reasons that today would be considered unjust, then there is a procedure for upgrading a dishonorable discharge. This is very important because, for example, a soldier might have been discharged for being gay, which is no longer considered justifiable grounds for such an action and certainly can be reversed. A good many veterans were separated with a dishonorable discharge for actions that could today be explained by PTSD. These are just a few examples.

Another detail worth remembering is that a dishonorable discharge as it relates to the uniformed services might not be so dishonorable in the eyes of the VA. What this means is that the Army administration might find a certain behavior and conditions unacceptable, but the VA might not necessarily regard that as disqualifying. Don't take anything at face value. *Don't forget the Five Ps!*

According to the Federal Guidelines, *"Even with a less than honorable discharge, you may be able to access some VA benefits through the Character of Discharge review process. When you apply for VA benefits, we'll review your record to determine if your service was 'honorable for VA purposes.'"*

This is an important subject, which, even though it affects a few individuals, is nonetheless a severe hindrance to those it affects, so that we will dwell on it in a little detail here. In general, when a member of the uniform services is discharged, the military assigns what is called a "character of service" on their DD-214, which, of course, is your all-important *Certificate of Release or Discharge from Active Duty.* Your discharge may be either:

- Honorable Discharge,
- General Discharge,
- Medical Discharge,
- Other than an Honorable Discharge,
- Bad Conduct Discharge, or
- Dishonorable Discharge.

Honorable Discharge

This is the top level of military discharge, qualifying a veteran for the full spectrum of VA benefits.

General Discharge

This means that, although a service member may not have met all of the standards of honorable discharge and may have some disciplinary marks on their record, their military service met standards of adequacy.

Other than an Honorable Discharge

This is when things start getting tricky. Other than honorable discharge, or a "bad paper," brings us into the category of undesirable discharge. An other than honorable discharge is an administrative discharge that is *not* determined by a court-martial. It complicates but does not render it impossible to obtain disability compensation and other veteran benefits.

Bad Conduct Discharge

This is a punitive discharge, and to be given one suggests some formal disciplinary procedure and a special court-martial determines this service character through a criminal trial.

Dishonorable Discharge

This classification is given to a service member who has committed a serious crime and is determined by court-martial. A dishonorable discharge renders a veteran ineligible for all VA benefits.

PRO TIP

Under the law, VA benefits are typically not payable to conscientious objectors who refused to perform military duty, a veteran who was discharged by general court-martial, who resigned as an officer for the good of the service, or who deserted. Benefits are also not payable to a veteran who was absent without official leave (AWOL) for a period of at least 180 days.

IMPORTANT

Unless it goes before a discharge review board for correction, your character of service determination remains on your military record, and even then this board may only modify, correct, or do discharge changes not imposed by a court-martial. If you are being denied VA benefits because of your discharge status, one option to consider is to apply for a discharge upgrade through the **Discharge Review Board (DRB)**. You may also appeal decisions from a discharge review board by submitting supporting documents and medical evidence if appropriate. The discharge review process is performed separately from and outside the VA. An important point is that even if a discharge is upgraded through the DRB, that does not automatically remove the bar to VA benefits.

Another very important detail to consider is whether your dishonorable discharge was related in some way to mental health – PTSD, for example – or some traumatic brain injury, sexual assault or sexual harassment, or, once again, sexual orientation because, under the Don't-Ask-Don't Tell policy, you have a more than excellent chance of securing a discharge upgrade. You can find a lot of information and advice on upgrading a dishonorable discharge by visiting www.va.gov/discharge-upgrade-instructions/.

The bottom line is that past military service qualifies you for veterans benefits; it's just a question of what you qualify for and can get.

How This Book Is Organized

ALRIGHT FRIENDS AND FELLOW VETERANS, let's get under the hood and have a look at what makes this thing run.

This book has been arranged to cover all of the bases from the ten thousand foot level and, wherever necessary, to point you in the direction of more detailed information. The VA is a huge, multi-faceted institution, organized or disorganized, some would say, on many levels, very often dissociated from one another. I am not trying to pack every available detail into this small publication, but simply to give you a general overview aimed at encouraging you to dig deeper and try harder to get what you deserve.

According to its information resources, the Veterans Administration offers: *"education opportunities and rehabilitation services and provides compensation payments for disabilities or death related to military service, home loan guaranties, pensions, burials, and health care that includes the services of nursing homes, clinics, and medical centers."*

As this brief assessment implies, the three main pillars of the VA benefits system are **pensions**, **healthcare**, including mental health resources, and **education**. Attached are multiple ancillary benefits such as VA housing and home loan services—which, for the sake of this book, we will combine with general financial services—job training and employment, and burials and memorials.

Each chapter is intended to provide a general overview of these principal pillars, with links and directions to where much more information is contained and many resources and tools to help you apply for and claim the benefits owed to you.

The biggest problem with dealing with the VA in particular, is if you are exploring some of the more obscure benefits, the massive bureaucratic edifice is characterized by all of the inefficiencies and exhausting red tape associated with any huge department.

It is not that the VA does not want to help and is not committed to doing so, but simply that the right hand often does not know what the left hand is doing and that many within the department are as frustrated with the protocols as those on the outside looking in.

The VA is one of the largest and most complex agencies in the US government, with an annual budget for fiscal year 2021 of $240 billion, and currently employs about 360,000 employees. It maintains and operates some 6,000 buildings, including 1,600 healthcare facilities, 144 medical centers, and 1,232 outpatient sites of differing complexity. The total number of veterans enrolled in the VA's health care system increased from 7.9 million to about 9.2 million from the fiscal year 2006 through the fiscal year 2022, and bearing that in mind, it is hardly surprising that dealing with the VA can sometimes feel like wading through molasses. Dealing with the VA is often described as a game of Chutes and Ladders, and according to Jim Vale, assistant director for claims at the American Legion – *"Veterans have heard stories, but many of them are surprised when they encounter the VA. The most important thing is to be represented."*

Well, that is certainly true, and there are a lot of specialists out there whose job it is to help you navigate the complexities of the system. Still, professional representation is expensive, and what you get from professional representation is often the same as what you would get from a little bit of private research and a lot of patience and persistence. This is not intended to be an exhaustive guide to VA benefits, and I will provide links to where you can find searchable specifics that will take you to where you need to go to deal with your specific query and help you jump through the hoops necessary to get what you are looking for. We aim to offer a broad overview of what's out there, how to navigate the system to find what you need, and a few tricks and tips to help.

IMPORTANT DOCUMENTS

Before you even get started, it is crucial to ensure that you have on hand your DD-214, DD-215, or, for World War II Veterans, your WD form. These documents are critical because they list service dates and types of discharge and are essential for any application for veteran's benefits.

Applying

The first thing you need to do is submit an intent to file. This sets the clock ticking, and no matter how long the process takes or when you actually file, all your benefits will be backdated to that point.

Applying for Health Benefits

There are different application processes for different benefit types, so let's kick off with the most important: health.

Usually, when you leave the service, you will be required to do a final out physical, which is not the time to be a hero. You are there to be evaluated for how broken you are. Typically, the physician conducting that process will be an active-duty doctor who knows the routine. Start with the big toe on your left foot and end with the hair follicles on the top of your head, listing every conceivable thing that is wrong, limited, or dysfunctional. **Stop moving when it starts hurting.** While the doctor will not encourage you to embellish, they should nonetheless encourage you to be expansive, and if they don't, do it anyway. All of it will stack up on your disability rating, and a good amount of what you end up with is based on that.

Ways to Apply

You can apply for VA health benefits online, via mail, telephone, or in person. Generally, I recommend submitting your application by mail. For more intel about this, check out www.benefits.va.gov/BENEFITS/Applying.asp.

These are the forms and documentation that you will need:

Firstly, you will need **VA Form 10-10EZ**, which you can go to www.va.gov/vaforms/medical/pdf/va_form_10-10ez.pdf in order to download the form. Then you will need the following supporting documents:

- Social Security numbers for you, your spouse, and your qualified dependents.
- Your military discharge papers (DD214 or other separation documents).
- Insurance card information for all insurance companies that cover you, including any coverage provided through a spouse or significant other. This includes Medicare, private insurance, or insurance from your employer.
- Gross household income from the previous calendar year for you, your spouse, and your dependents. This includes income from a job and any other sources. Gross household income is your income before taxes and any other deductions.
- Your deductible expenses for the past year. These include certain health care and education costs.

NOTE

It is unnecessary to reveal any details about your income and expenses upon application. However, should you be ineligible for other reasons, it might be necessary to provide that information for your application to be processed.

Having submitted your application, it is a question of waiting. My application was decided in ninety days, and that is about the average wait time. Usually, you will receive an acknowledgment of receipt of your application with the information included in any follow-up or query.

Applying for the GI Bill

The process of applying for education benefits is similar to health, although the only documentation that you need are your separation forms and:

- Social Security number,
- Bank account direct deposit information,
- Education and military history, and
- Basic information about the school or training facility you want to attend or are attending now.

You can apply online, in person, through the mail, or with a professional agent. Generally, you will be looking at a processing time between thirty and ninety days, usually closer to the former. At least seeking the advice of your local state VA is a great start.

Healthcare

"For years, I thought the VA was for only combat-wounded veterans, and it took me over twenty years to figure out that this was not true. Then, I began strategically entering my medical conditions without getting 'flagged' as 'lame, lazy, or crazy.' I wanted to be ready and available for deployment and not placed on some disqualifying medical status."

-Lt. Col. Jason G. Pike, USA, Retired

IN THIS CHAPTER, WE WILL LOOK AT WHAT IS ARGUABLY the most important pillar of the veterans' benefit system: its healthcare system.

Firstly, I'll deal with the greatest misconception among ex-servicemen and veterans: that veterans' physical and mental healthcare is reserved for those men and women who served on the frontline in a combat capacity and who suffered some related trauma or severe physical injury. This is not true! Veterans' benefits across the spectrum are available to all men and women who wore the uniform—*accepting what is due to you as a veteran does not take away from anybody else!* There are, however, a few eligibility factors that need to be taken into consideration, so let's deal with those first.

PRO TIP

Keep all your medical records and any paperwork related to health and medical issues for the entire time you are on active duty. When it comes time to submit your claim for VA health benefits and to establish your priority group and disability level, these will be invaluable.

Eligibility

Now, let's take a look at the question of eligibility.

We have already touched on the essential standards of eligibility for veterans benefits. Still, as it relates to healthcare, it is worth reiterating that unless you were dishonorably discharged, you are eligible for subsidized healthcare through the Veterans Health Administration if you served twenty-four consecutive months or completed the entire period of active service you were called for. If you were discharged because of a service-related disability or served before September 7, 1980, you are not obligated to meet the minimum active duty requirements. **There are minor points of disparity here,** and you can research those by visiting www.va.gov/health-care/eligibility/, but basically, if you satisfy the above criteria, then you are in.

Besides these general eligibility criteria, all you need to concern yourself about regarding health benefits is the potential that you might receive *enhanced* benefits if at least one of these is true:

- You receive financial compensation (payments) from the VA for a service-connected disability,
- You were discharged for a disability resulting from something that happened to you in the line of duty,
- You were discharged for a disability that got worse in the line of duty,
- You're a combat Veteran discharged or released on or after September 11, 2001,
- You get a VA pension,
- You're a former prisoner of war (POW),
- You have received a Purple Heart,
- You have received a Medal of Honor,
- You get (or qualify for) Medicaid benefits,
- You served in Southwest Asia during the Gulf War between August 2, 1990, and November 11, 1998, and/or
- You served at least 30 days at Camp Lejeune between August 1, 1953, and December 31, 1987.

Or, you must have served in any of these locations during the Vietnam War era:

- Any US or Royal Thai military base in Thailand from January 9, 1962, through June 30, 1976,
- Laos from December 1, 1965, through September 30, 1969,
- Cambodia at Mimot or Krek, Kampong Cham Province from April 16, 1969, through April 30, 1969,
- Guam or American Samoa or in the territorial waters off Guam or American Samoa from January 9, 1962, through July 31, 1980,
- Johnston Atoll or on a ship that was called Johnston Atoll from January 1, 1972, through September 30, 1977, and/or
- The Republic of Vietnam from January 9, 1962, through May 7, 1975.

PRO TIP

I have spoken to vets who take the view that even with a dishonorable discharge, the VA is obligated to provide health benefits. This is a contestable point but one worth bearing in mind.

The job of the Veterans Health Administration (VHA), the largest health system in the United States (which makes it one of the largest in the world), is to provide eligible veterans with comprehensive medical care. While it has many branches and contributes much to medical

---

research, its primary function is to take care of you, a United States military service veteran.

Healthcare benefits are, to some extent, available to almost all US military veterans, although they might not always be absolutely free. While you will be eligible for free care for any specific injuries, illness, or ailments that are in some way service-related, completely free general healthcare is only available to veterans with a disability rating of at least fifty percent.

Now, the question of disability ratings is key to the process of assigning healthcare to military veterans, and again, this is not contingent on battlefield injuries. Bearing in mind the extensive range of conditions and situations that qualify as a disability in this context, establishing yourself as fifty percent or more disabled is not that difficult.

You can find a range of information about this here:
www.va.gov/disability/about-disability-ratings/.

According to Federal Guidelines, both of these must be true:

- You have a current illness or injury (known as a condition) that affects your mind or body, and

- You served on active duty, active duty for training, or inactive duty training.

This is a good start because you would not be reading this book if you had no association with the military, so we can take the latter point as a given. With regards to the nature of your "condition," according to the federal guidelines, see below.

At least one of these must be true:

- You got sick or injured while serving in the military—and can link this condition to your illness or injury (called an in-service disability claim),
- You had an illness or injury before you joined the military—and serving made it worse (called a pre-service disability claim), or
- You have a disability related to your active-duty service that didn't appear until after you ended your service (called a post-service disability claim).

Again, none of this relates necessarily to specific combat-related injuries or disabilities. Issues can range from simple range-of-motion limitations to hearing loss, back pain, breathing difficulties, and any number of stress-related conditions, with PTSD being the most obvious. If you suffer from such conditions as Gulf War Syndrome, respiratory difficulties associated with burn pits, and radiation poisoning (or, in fact, just about any poisoning), your disability rating is likely to be high. One hundred percent disability is not uncommon. Hell, I have a one hundred percent disability, and I never fired a shot in anger.

The idea of one hundred percent disability implies to most people that you are dead, or at least as good as, but that is not the case in the VA universe. Don't ask me how the VA rationalizes its disability ratings; it is just one of those great mysteries, but it is important to understand that gaining a fifty-percent plus rating is not that difficult and will qualify you for free healthcare under the VA system. It is also probably worth noting that free VA healthcare is also usually available automatically for indigent or destitute veterans.

Meeting the Unique Needs of Women Veterans

Local VA facilities offer a variety of services, including women's gender-specific health, screening, and disease prevention, maternity care, and routine gynecologic services. LGBTQ+ Veterans are eligible for the same VA benefits as any other veteran. **Check out this section on LGBTQ+ and Women's issues on the VA website here:** www.patientcare.va.gov/lgbt/.

Another point worth noting is that free healthcare for dependents is only available for those with a one-hundred percent disability rating, and this is typically through the Civilian Health and Medical Program of the Department of Veterans Affairs, more commonly known as CHAMPVA. We will touch on this in more detail later.

KEY TAKEAWAY

The main thing to remember is to work towards a disability rating of at least fifty percent. This is no time to be a hero. Even if you polished your ass on a desk chair for your entire career, you wore the uniform, served, and qualified. Now let's get into **Priority Groups**, another important aspect of VA healthcare access.

Priority Groups

This detail of the VHA system is well worth familiarizing yourself with. Priority Groups are basically a standard for gauging the urgency of individual needs based on certain criteria. As of 2021, there were 16.5 million veterans alive in the United States, and, obviously, it is impossible to offer free and unlimited care to all, so it becomes necessary to prioritize those with the greatest need.

According to the Federal Guidelines, *"When you apply for VA health care, we'll assign you to 1 of 8 priority groups. This system helps ensure that Veterans who need care right away can get signed up quickly. It also helps to make sure we can provide high-quality care to all Veterans enrolled in the VA health care program."*

Your priority group will be determined and assigned according to your military service history, your disability rating, *and* your income level, whether or not you qualify for Medicaid, *and* what other benefits you may be receiving.

As a general rule of thumb, you will be assigned **Priority 1** if you have an above fifty-percent disability rating, are unable to work, and/or are a Medal of Honor awardee. Criteria diminish until, at the **Priority 8** level, you have a minimal disability, earn above the VA income limits, and are willing to pay copays. The VA considers a variety of different factors when assigning priority groups, and obviously, the higher your priority group, the sooner you'll get your benefits. Your priority group also influences how much, if anything at all, you will be required to pay in copays. As a rule, priority groups seven and eight have copays associated with benefits.

Go to www.va.gov/health-care/eligibility/priority-groups/ for much more detailed information about priority groups, what the specific criteria are, and how that might affect your application.

Like your disability rating, your priority group is a joyous work in progress that culminates only on the day that you need it no more. It is always good to keep on top of both and get them adjusted and amended from time to time.

Healthcare for Spouses and Dependents

For many vets out there, healthcare for spouses and dependents is a key aspect of VA benefits. In fact, family members can benefit not just in the area of healthcare. There are opportunities for education benefits, too, and we'll get into that a little bit later. For the moment, let's get into CHAMPVA, or *Champ-VA*, as it is better known.

According to the Federal Guidelines, *"the Civilian Health and Medical Program of the Department of Veterans Affairs (CHAMPVA) is a health benefits program in which the Department of Veterans Affairs (VA) shares the cost of certain health care services and supplies with eligible beneficiaries."*

Who are eligible beneficiaries? CHAMPVA is a health-care option for qualifying veteran spouses, children, stepchildren, and other dependents. If you are 100 percent disabled, then your spouse and dependents will most likely be covered by CHAMPVA. There are caveats and particularities as, of course, there always are, and one that almost goes without saying is that if you were dishonorably discharged, then your dependents are ineligible. In general, however, under the normal rules of eligibility, your dependents should be covered by CHAMPVA.

You will find a lot more information on CHAMPVA here: **www.va.gov/health-care/family-caregiver-benefits/champva/.**

Veteran Dental Care

Subject to the same essential eligibility standards of all veterans' healthcare benefits, dental care is available to most. As it concerns dental care, the question of family members is a bit more complicated and needs to be looked into carefully. You will also find a lot more information about that by visiting www.va.gov/health-care/about-va-health-benefits/dental-care/.

Reimbursement of Travel Costs

Eligible veterans and non-veterans may be provided mileage reimbursement or, when medically indi-cated, special mode transport (e.g., wheelchair van, ambulance) when travel is in relation to VA medical care.

Va Education Benefits

ANYONE READING THIS WHO HAS READ MY MEMOIR will know that I put on the uniform of the US National Guard primarily to help fund a college education. My Dad did the same thing, following a route forged by a great many other young men of the South since the GI Bill began. Through my military career I've earned two master's degrees and my daughter has just gotten her bachelors, and is now looking toward graduate school. Both of our higher education goals have been free so far. What the GI Bill has done for me and my family is impossible to put a figure on.

In my opinion, the GI Bill defines the attitude of the government and general public of this nation toward their military veterans. Nothing speaks more clearly of society's appreciation than offering a returning soldier access to education. It uplifts those with a will to succeed, and it enriches society. It adds to the professional ranks of the nation a cadre of men and women with practical skills, an understanding of the chain of command, and the benefit of training at the hands of arguably the greatest military structure in the world.

The GI Bill came into existence as the Servicemen's Readjustment Act of 1944, and it was established initially to encourage enlistment. Since then, it has become the very cornerstone of the VA benefit system. The original idea was conceived and driven by the American Legion and then signed into law by President Franklin D. Roosevelt. Although the original intent of the GI Bill was to offer mainly education benefits, in the end it grew into an umbrella term for a whole range of allied benefits. Those included low-cost mortgages, low-interest loans to start a business or farm, one year of unemployment compensation, and dedicated payments of tuition and living expenses to attend high school, college, or vocational school. These were available to all veterans who had been on active duty during the war years and, of course, had *not* been dishonorably discharged.

Before its expiry in July 1956, eight million veterans received educational benefits as part of the GI Bill. The number of degrees awarded by US colleges and universities more than doubled between 1940 and 1950, and the percentage of Americans with bachelor degrees or advanced degrees rose from 4.6 percent in 1945 to twenty-five percent a half-century later. Over the years, it has made a significant contribution to the quality of US human capital and has been a major factor in the long-term, post-war economic growth of the United States.

There have been quite a few variations and updates to the GI Bill, and we do not really need to go into any detail about that, perhaps to say that in the twenty-first century, the "GI Bill" is alive and well and filled with bounty for an ambitious ex-member.

If you are a service academy graduate—Air Force Academy, West Point, Annapolis, or Coast Guard—*you will not be eligible* for the Post-9/11 GI Bill benefits unless you served extra time on top of your five-year active duty service requirement.

Eligibility

Basically, if you are a Purple Heart recipient on or after September 11, 2001, served on active duty for at least thirty days continuously, or were discharged with a service-connected disability, then you are eligible. Again, if you were dishonorably discharged, you are not. The amount of money that you receive will depend on how long you serve. At least ninety days of active-duty service since September 10, 2001, is required to receive fifty percent of tuition costs, after which the VA prorates the benefit up to the capped amount.

You can take the following as a rule of thumb:

- Ninety days to five months (90-179 days), fifty percent of the max benefit,

- Six to seventeen months (180-544 days, sixty percent of the max benefit,
- Eighteen to twenty-three months (545-729 days), seventy percent of the max benefit,
- Twenty-four to twenty-nine months (703-909 days), eighty percent of the max benefit,
- Thirty to thirty-five months (910-1,094 days), ninety percent of the max benefit, and
- Thirty-six months (1,095 days or more), one hundred percent of the max benefit.

The Post-9/11 GI Bill is not the only education benefit system, but it is undoubtedly the most comprehensive, and if you qualify for it, you should certainly make the most of it. If, on the other hand, you served *before* September 10, 2001, then you might consider the Montgomery GI Bill Active Duty and the Montgomery GI Bill Selected Reserve. Bear in mind that once you have made your choice, it is irrevocable, so if you are interested in something other than the Post-9/11 GI Bill, make sure you have applied the Five Ps.

You will find more detailed information on all of this here: www.va.gov/education/.

The Yellow Ribbon Program

What is the Yellow Ribbon Program? **According to the VA website:** *"The Yellow Ribbon Program can help you pay for higher out-of-state, private school, foreign school,* *or graduate school tuition and fees that the Post-9/11 GI Bill doesn't cover. Keep reading to find out if you're eligible and if your school takes part in this program."*

In more general terms, the **Yellow Ribbon Program** is a provision of the Post 9/11 GI Bill, specifically aimed at helping students attend expensive private schools or private or out-of-state colleges at little or no cost to themselves. Typically, Post-9/11 GI Bill payments to such institutions are limited to a national maximum amount allowed by law. Although that amount tends to fluctuate yearly, it only sometimes covers the full tuition and fees that private schools charge for enrollment.

Under the **Yellow Ribbon Program**, schools can voluntarily enter into an agreement with the VA to waive some or all of their tuition costs that exceed the national maximum Post-9/11 GI Bill reimbursement, and the VA will match the amount of the waiver. If, for example, a student wishes to attend a private university that charges an annual tuition fee of $50,000 per year, and the school agrees to waive $10,000 tuition for Yellow Ribbon Participants, the VA will match that $10,000. This means that the total tuition waived is $20,000. Your Post-9/11 GI Bill grant will cover $26,381.37 (the maximum at the time of writing), and the Yellow Ribbon Program will waive $20,000, leaving you with a total effective grant of $46,381.37. You will be responsible for the $3,618.63 difference you must pay with financial aid or out-of-pocket.

Eligibility to Receive the Yellow Ribbon Benefits

You must obviously qualify first for the Post-9/11 GI Bill at the one hundred percent benefit level, after which the general VA qualifications apply. You can also qualify as a spouse using the transferred benefits of an active-duty service member who has served at least 36 months on active duty, or a dependent child using benefits transferred by a Veteran, or you are a Fry Scholar (The Marine Gunnery Sergeant John David Fry Scholarship provides scholarships for children and spouses of certain Veterans). Also, your proposed school must be a party to the Yellow Ribbon Program.

You will find a whole lot more information about the Yellow Ribbon Program by visiting www.va.gov/education/about-gi-bill-benefits/post-9-11/yellow-ribbon-program/.

Dependents and Survivors

The beauty of the GI Bill education benefit system is that spouses, children, and even stepchildren and surviving dependents may qualify for education benefits in many instances. As I have already mentioned, my daughter, Chantel, received the Post 9/11 GI Bill and a four-year free education.

How does this work? The specific program is called the Survivors' and Dependents' Educational Assistance

(DEA) program. According to the VA: *'if you're the child or spouse of a Veteran or service member who has died, is captured or missing, or has disabilities, you may be able to get help paying for school or job training through the DEA program—also called Chapter 35."*

PRO TIP

The main caveat to the DEA program is that it is available only if you have served, or intend to serve, for a ten-year minimum. In effect, this means that it is unnecessary to serve the entire ten years of your commitment before the benefits become available. Applications can be submitted after at least six years, with a commitment for a further minimum of four years.

VA Home Loan And Housing Benefits

AN OFTEN OVERLOOKED, BUT HUGELY BENEFI-CIAL VA BENEFIT is the VA Home Loan Program, which offers veterans the opportunity to purchase a home without having to deal with many of the pitfalls and expenses of dealing with a private lender.

According to the Federal guidelines, *"the VA helps Veterans, Servicemembers, and eligible surviving spouses become homeowners. As part of our mission to serve you, we provide a home loan guarantee benefit and other housing-related programs to help you buy, build, repair, retain, or adapt a home for your own personal occupancy."*

Here are a few of the signature benefits of the VA home loan benefit:

- No down payment required (Lenders may require down payments for some borrowers using the VA home loan guaranty, but VA does not require a downpayment),
- Competitive interest rates,
- Limited closing costs, and
- No requirement for Private Mortgage Insurance (PMI).

PRO TIP

Something worth remembering is that the VA home loan is a lifetime benefit, and assuming you are in good standing, you can utilize the guarantee multiple times.

Eligibility

Typically, if you are either an active duty service member or a veteran, you will likely qualify for the VA Home Loan Program. There are a few things, however, that you need to know. On the understanding that disabled vets will form a large part of the demographic utilizing this benefit, the VA does offer additional grants, subject to a few conditions, to adapt your home to your specific needs. Obvious examples would be wheelchair accessibility and chair lifts and such. You will need to check out the federal guide-lines for more information on that. Also, if you qualify under this particular rule, the VA offers loans directly to members of Native American groups, usually on much better terms than private lenders. Once again, those with a dishonorable discharge do not qual-ify. Certain eligible spouses and other uniformed service personnel may be eligible for VA home loan guaranty benefits. You will need to check in with the *VA Home Loan Guaranty Buyer's Guide* on the VA website for more detailed guidelines, but here are some of the basics.

For a whole lot more intel on this benefit, visit www.benefits.va.gov/homeloans/.

The Keywords in This Section are "Loan Guaranty"

In most cases, the VA does not lend money directly but guarantees loans from private lenders. Before you can move forward, however, a Certificate of Eligibility, or COE, is required for a private lender to use secure VA backing. To acquire a COE, you will need to meet minimal active duty service requirements, the minute details of which are available in the *VA Home Loan Guaranty Buyer's Guide* or visit here - www.va.gov/housing-assistance/home-loans/how-to-request-coe/. These are, however, generally the same basic qualifications that got you into the general VA benefits. Active duty National Guard and Reserve members typically qualify, and certain spouses also, but likewise under certain, specific conditions.

Many private lenders are working exclusively in the VA loan market, and it is obviously advisable to search a few out in favor of traditional mortgage lenders because they are familiar with the routine and more sensitive to your specific needs and requirements. Another very important point is that the VA does not have a minimum credit score requirement, which is huge, although private lenders still do, so you will need to meet both their credit score and income requirements to nail down a loan, even if the VA backs it up. You may still qualify for a VA home loan if you have poor credit, but the terms and options might be more restrictive.

Yet another relevant point is that while securing a home loan through the VA will save tens of thousands of dollars in the long run, there is a service charge, usually calculated at 1.4 percent and 3.6 percent of the full borrowed amount. However, if you have a service-related disability rated higher than ten percent, likely the VA will waive that fee. If you go through the process and pay that funding fee, and then later a disability rating upwards of ten percent, then that fee will be refunded.

TRANSITIONING INTO CIVILIAN LIFE

Military service always tends to look very good on an individual's resume, and assuming that you do not have a dishonorable discharge, most employers will look very favorably on a job application from a military veteran. In the military, you are trained to be a team player, to be mission-focused, and to respect and appreciate a chain of command. That is something that potential employers love. However, to back that up, the VA offers a whole spectrum of programs aimed at helping veterans secure and maintain good employment.

Every veteran transitioning out of uniform will surely agree that moving from the institutionalized world of the military, where every task is assigned, into the less regulated and more individualistic world of civilian employment can sometimes be very difficult. This is particularly true if you have a disability rating. In preparing this program, I have spoken to a lot of vets, and in almost every case I have heard that the transition can often be very difficult. Civilian organizations very rarely run along the clear lines of a linear chain of command, and an ex-serviceman who is accustomed to the clarity and unambiguousness of the military command structure can often grow confused and frustrated at the many-headed dragon that is the corporate hierarchy.

There are many VA programs – and, in fact, quite a number that are not VA-run, but nonprofits and foundations that have been set up to help and guide veterans through this process. These can take the form of workshops and seminars, formal curriculum, training grants, and contact groups, as well as extensions to GI Bill education benefits and various vocational training. The bottom line is that there is absolutely no shortage of assistance programs for vets to plug into. Many of these are state run programs, and you will find much more detailed information at your particular state VA office, including telephone numbers and email addresses and a constantly updated list of non-VA affiliate programs offered by the many non-profits out there set up to help.

Here we will mention just two cornerstone programs available to veterans through the Veterans Administration.

The first of these is the **Veterans' Preference**. This is not a benefits program as such but simply an understanding that veterans will be given preferential consideration when applying for employment with any federal agency. **According to the Department of Labor:** *"veterans who are disabled, who served on active duty in the Armed Forces during certain specified time periods or in military campaigns are entitled to preference over others in hiring for virtually all federal government jobs."*

PRO TIP

Certain service members and veterans might be eligible for a one-time payment toward purchasing a motor vehicle if they have certain service-connected disabilities. Likewise, a veteran with a service-related disability who requires prosthetic or orthopedic appliances may be eligible for a clothing allowance.

Veteran Readiness and Employment (VR&E)

According to the Federal Guidelines, *"if you're a Veteran or service member with a service-connected disability that impacts your ability to work, the Veteran Readiness and Employment program (formerly called Vocational Rehabilitation and Employment) may be able to help. We offer five support-and-services tracks to help you get an education or training, find and keep a job, and live as independently as possible."*

PRO TIP

As a Veteran, you're protected under the Uniformed Services Employment and Re-employment Rights Act (USERRA). This means you can't be disadvantaged in your civilian career because of your service.

The five support-and-service tracks offered by the VA include:

1. **Reemployment track:** If you're a Veteran with a service-connected disability, the reemployment track can help your employer accommodate your needs. Your Vocational Rehabilitation Counselor (VRC) can provide a full range of services and can refer you directly to the Department of Labor to begin the process.

2. **Rapid Access to Employment track:** If you want to follow an employment path that uses your existing skill set, the Rapid Access to Employment track can help you with your job search. They offer counseling and rehabilitation services that address your abilities, aptitudes, and interests.

3. **Self-Employment track:** If you're a service member or Veteran with a service-connected disability and employment barrier who has the strong desire, skills, and drive to run a successful business, you may be interested in the Self-Employment track.

4. **Employment Through Long-Term Services track:** If you have a service-connected disability that makes it hard for you to succeed in your employment path, you may be interested in the Employment Through Long-Term Services track. They can help you get the education or training you need to find work in a different field that better suits your current abilities and interests.

5. **VR&E Independent Living track:** If your service-connected disability limits your ability to perform activities of daily living (like bathing, dressing, accessing the community, and interacting with others) and you can't return to work right away, you may qualify for independent living services through the Independent Living track. You may also receive these services as you work to find a job if that's a goal you and your Vocational Rehabilitation Counselor (VRC) have created. In both cases, your VRC can help you restore your daily living activities.

PRO TIP

If you are a reservist and are called back to duty, you can return to your old job at the end of your deployment based on certain conditions. If you are a full-time employee, the employer is mandated to hold the position for you. They also cannot withdraw any benefits based on seniority, including pay increases and promotions.

Moving On: VA Burial & Memorials

Although they say that old soldiers never die, they do, and the VA is there to help with final arrangements. Help, however, is the keyword. The VA will not cover the cost of *all* the expenses associated with your funeral, and, at most, your family will get $2,000, with the possible addition of costs related to the transport of your remains. That is valuable assistance for most people, and here is how to get it.

Service-Related Death

The VA will pay up to $2,000 toward burial expenses for deaths on or after September 11, 2001, or up to $1,500 for deaths before September 11, 2001. If the Veteran is buried in a VA national cemetery, some or all of the cost of transporting the deceased may be reimbursed.

Non-Service-Related Death

The VA will pay up to $796 toward burial and funeral expenses for deaths on or after October 1, 2019 (if hospitalized by VA at the time of death), or $300 toward burial and funeral expenses (if not hospitalized by VA at the time of death), and a $796 plot-interment allowance (if not buried in a national cemetery). For deaths on or after December 1, 2001, but before October 1, 2011, the VA will pay up to $300 toward burial and funeral expenses and a $300 plot-interment allowance. For deaths on or after April 1, 1988, but before October 1, 2011, the VA will pay $300 toward burial and funeral expenses (for veterans hospitalized by the VA at the time of death).

PRO TIP

Of the 136 currently operating national cemeteries, almost 100 of them are accepting new internments. If you met active duty service requirements and were discharged under conditions other than dishonorable, you are welcome. In some cases, surviving members of your family may also be eligible for burial in a national cemetery. Further, the VA will make and deliver a headstone or marker anywhere in the world, along with a burial flag. The good news is that you can figure out all these logistics now instead of leaving the work to your surviving family members. That's what the VA's Pre-Need Burial Eligibility Determination program is for.

Mental Health Care

TWENTY-YEARS AFTER THE START OF THE GULF WAR, more vets have died by suicide than fell in action. According to a study by the Watson Institute of Brown University, **30,177 active duty personnel and veterans of the post-9/11 wars have died by suicide**, a figure four times higher than the 70,757 service members killed in combat. Suicide is a silent epidemic, far more deadly than combat, and if this book achieves anything, I hope it will be to persuade a vet on the brink to go and get help. Dial 988 for the Veteran Crisis Line, and then press one, and you will be connected to a sympathetic and qualified responder. It might make all the difference.

You can also access emergency mental health care twenty-four-seven at any VA Medical Center. If not there, then every VAMC is obligated to offer such care through other appropriate facilities. Care is always available; outreach is always available, so make use of it!

PRO TIP

Regardless of your discharge status (including dishonorable), service history, or eligibility for VA healthcare, you are eligible for free mental health care for a year after your separation.

How to Access Mental Health Resources

- Call or visit any VA medical center 24/7, 365 days a year!

You will find a comprehensive list of facilities here - www.va.gov/find-locations.

- Call or visit one of more than two hundred different Vet Centers across America during clinic hours for readjustment counseling.

Again, visit www.va.gov/find-locations for a complete list of locations.

- Call 1-877-222-8387 Monday through Friday, 8:00 a.m. to 8:00 p.m. ET, for help finding the right resources suited to your needs.
- At any time, you can call the Veterans Crisis Line at 1-800-273-8255 (and then press 1) or text 838255.

DON'T BE A STATISTIC! GET HELP!

State Benefits

WE HAVE ALREADY BRIEFLY SPOKEN ABOUT state-level Veterans Administration regarding seeking help in the application process. I will reiterate that state VA offices should be your first call when approaching the application process. State VA representatives are salaried public servants whose job it is to help. Individual state VA offices are not the huge, monolithic edifices that the federal VA is, so your interface will be more personal, quicker, and more efficient.

Another point worth bearing in mind is that state VAs, depending on the state, offer a whole additional cornucopia of benefits on top of federal VA benefits. Some vets I spoke to chose which state to settle in based on what local and state VAs had to offer.

PRO TIP

State benefits range from free college and employment resources to free hunting and fishing licenses. Most states also offer tax breaks for their veterans and specialized license plates; some even provide their veterans with cash bonuses just for serving in the military.

KEEP ON TOP OF YOUR STATE BENEFITS. THEY ARE ALTERED AND CHANGED REGULARLY.

A Whole World Of Opportunities

NOW, WE COME TO THE UNIVERSE OF PROGRAMS AND OPPORTUNITIES offered by organizations, trusts, and nonprofits set up for veterans by veterans, and the list, literally, is almost endless. Many businesses offer veterans discounts, and most national and state parks offer free parking. These are two of many opportunities out there that are available to vets. Some programs are general while others are specific to individual units or detachments and even to particular wars. The vast preponderance, however, are open to all vets and their families and can range from clubs and associations to shared events to education benefits to aid and help in a crisis. *There are far too many to list here, so again, check out my website at JasonPike.org for a comprehensive and constantly updated list of opportunities.*

All of this adds up to the fact that the VA is there to help and has many offers, so get proactive and claim what is yours TODAY!

Dear fellow veterans,

Congratulations on completing this comprehensive guide to accessing VA benefits and assistance programs after getting out of the uniform and back into civilian life! As a fellow veteran who has walked the same path, I understand the challenges and questions that can arise during this journey. But remember, we are not alone in this pursuit of the support and recognition we rightfully deserve.

Let's stand strong together, ensuring every Veteran gets the benefits and assistance they've earned through their service. You've fought for our nation; now it's time to ensure you receive the recognition and support you deserve.

To connect with a community of veterans who understand your journey and can provide support, camaraderie, and a shared commitment to ensuring no one gets left behind, follow me on Instagram @AuthorJasonPike and Facebook @JasonPike, or visit my website www.JasonPike.org

Thank you for your service, and welcome to our veteran family!

Lt. Col. Jason G. Pike, USA, Retired
www.JasonPike.org

About The Author

Adecorated combat veteran with multiple deploy-ments, Lt. Col. Jason G. Pike, USA, Retired, served 31 years in the United States Army as both an enlisted and officer, including nine years overseas in five coun-tries. Jason earned over 30 service awards & badges and survived a wicked amount of military training.

His first book, **A Soldier Against All Odds**, compiles all his life events in an inspiring storytelling format with the ups and downs of a life in uniform. His diversity of Army jobs, assignments, and schools from age 17 to 48 sets this military memoir up differently than most.

Jason's brutal honesty on how he did it while disclosing many sacred secrets about how he survived is unique. With a straightforward account of one man's journey, he inspires audiences nationwide at speaking events and shows how to be resilient and to persevere no matter what disadvantages and life struggles may happen.

After having walked through the VA benefits bureaucracy and endless paperwork, Pike has made it his business to master getting the benefits he earned, and he now has a much-needed blueprint to help other veterans do the same. His second book, **Out of the Uniform, Back into Civilian Life**, is an invaluable resource for veterans seeking clear, actionable guidance to navigate the often complex landscape of VA benefits and assistance programs.

ACKNOWLEDGEMENTS

On the publishing side, thank you to my amazing team. To my lead editor, Peter Baxter, for your insight, expertise, and guidance throughout the entire process of creating the book. Also, thanks to Bryan Davis and Cody Rollins for your perfectionism and keen eye.

Thank you to my publishing consultant, PRESStinely—to Kristen Wise, Maira Pedierra, and their amazing staff. I could not have done this without you.

To my family, you are my rock and my inspiration—God only knows where I would be without each of you.

And to you, the reader and fellow veteran: Thank you for your service and sacrifice in protecting our freedoms we so often take for granted. We have a brighter future because of your commitment to our country.